YOUR KNOWLEDGE HAS VALUE

- We will publish your bachelor's and
  master's thesis, essays and papers

- Your own eBook and book -
  sold worldwide in all relevant shops

- Earn money with each sale

Upload your text at www.GRIN.com
and publish for free

Bibliographic information published by the German National Library:

The German National Library lists this publication in the National Bibliography; detailed bibliographic data are available on the Internet at http://dnb.dnb.de .

Imprint:

Copyright © 2015 GRIN Verlag, Open Publishing GmbH
Print and binding: Books on Demand GmbH, Norderstedt Germany
ISBN: 978-3-668-04647-4

This book at GRIN:

http://www.grin.com/en/e-book/306848/design-and-fabrication-of-groundnut-shelling-machine

S.O. Ejiko, J.T. Adu, Osayomi Peters

Design and Fabrication of Groundnut Shelling Machine

GRIN Publishing

DESIGN AND FABRICATION OF GROUNDNUT SHELLING MACHINE

[*1]Ejiko, S.O, [2]Adu, J.T. and [3]Osayomi, P.

[1, 3]Department of Mechanical Engineering, The Federal Polytechnic, Ado-Ekiti
[2]Department of Mechanical Engineering, Obafemi AwolowoUniversity, Ile-Ife, Osun State, Nigeria
jcdatjiko@yahoo.com; jamtolad@gmail.com; osayomipeters@yahoo.com

ABSTRACT

Groundnut product demand is on the increase and the application is largely dependent on the cleanness of the nuts. The separation process is usually an energy sapping task that requires a lot of time. In order to separate the nuts from its shell effectively a shelling machine was developed. The machine employs an auger screw as a means of breaking the groundnut pod. The machine basically comprises of shelling chamber, separating chamber and a motor (1HP). The arrangement of these parts is connected by a compound belt of type B standard V-belt of pitch length 1694mm. With the Von-mises equation, the material for the shelling shaft is taken to be mild steel. The materials used in the fabrication of the machine are sourced locally so as to ensure that it is cheap, affordable and easily maintained by the peasant farmers. The shelling efficiency and material damage are 84% and 14% respectively for groundnut seeds of 86.5% dry.

Keywords*: Groundnut, pods, shelling machine, design methodology, efficiency, fabrication*

TABLE OF CONTENTS

1.0 INTRODUCTION

Groundnut (Arachis Hypogea) originated from latin America (Brazil) and was introduced into West Africa by Portuguese traders in the 16th century. Its actual origin dates back to 350BC (Komolafe *et al.*, 1985 and Hommons, 1994). It is botanically a member of the papilionaceae and is the largest and most important member of the leguminosae (Shankarappa, Robert and Virginia, 2003). This crop is considered to have great potential for both food and industrial purposes in the tropical regions of Africa (Milner, 1973). In some parts of Nigeria like Kano, Sokoto, Kaduna, Bauchi and Borno States, it is used as a source of revenue generation to support rural development (Maduako *et al.*, 2006).

One of the important processes involved in the production of groundnut is shelling and separation. Shelling is the removal of the groundnut seed from its pod by impact action, compression and shearing or combination of two/more of these methods. The shelling operation is majorly divided into two, namely: traditional and mechanical methods. The traditional shelling could be by stick beating, animal trampling or pod pressing by hand. Pod pressing is mostly practiced in Nigeria and it has low efficiency, high energy requirement, time wastage, high labour and fatigue.

Several shelling machines based on different techniques have been developed over the years. Among these machines are hand-operated groundnut sheller, baby groundnut sheller, powered rubber groundnut sheller to mention but few (Rodriquez, 1985). Due to custom charges, these machines are very expensive and are not affordable by the small-scale producers and farmers.

Maduako and Hamman (2004) determined some physical properties of three groundnut varieties. The materials used were ICGV-SM-93523, RMP-9 and RMP-12. Some of the physical properties considered were weight, bulk density, moisture content etc. Anova was used for analysis at 0.5 probability level.

Okegbile *et al.*, (2014) designed and fabricated groundnut shelling and separating machine. With 1hp motor, the machine has shelling capacity of 400kg/hour and shelling efficiency of 78% and material damage is 17.25%. The materials used were sought locally. The machine used spike to crush the pods.

This research aimed at using anger shaft in the crushing of the pod to replace the existing design. Also, to improve on the shelling capacity at low cost by using locally – available materials. Then, the machine is evaluated based on shelling efficiency and percentage material.

2.0 MATERIALS AND METHODS

2.1 Materials

The materials used were locally obtained from a town called Ado-Ekiti which is the capital of Ekiti State in Nigeria. The selection of these materials was based on durability, cost and availability, strength and rigidity, weight and friction.

2.2 Description of the Groundnut Shelling Machine

Various components were put together in the fabrication of this machine. These parts are: the frame, hopper, seed discharge outlet, shelling chamber, fan and chaff outlet.

Frame

It holds the hopper, shelling, and separating unit as well as the prime mover (electric motor). Being the main support for the machine, it must be able to withstand stresses and loads and have good welding properties. Hence, mild steel in form of angle bar was used.

Hopper

It contains the unshelled groundnut before and during the shelling operation. It must be able to withstand the vibration loads and stresses, have good strength and good corrosion resistance. Hence, the material is mild steel sheet of 2mm thickness.

Shelling Chamber

It houses the auger and the shelling drum. The shelling operation is done inside it. Therefore, it must be able to withstand load and stresses, good weld ability and corrosion resistance. The diameter of the shelling drum is 206mm and pitch of the anger screw is 100mm.The active length of the drum is 500mm. So, mild steel of 2mm thickness was selected.

Seed Discharge Outlet

The shelled groundnut seed is collected through this outlet. The seeds fall under gravity from the shelling chamber into its tray. It must have good strength and high resistance to impact loads. So, mild steel of 2mm thickness was used.

Chaff Outlet

The broken pod is separated from the groundnut by pressure provided from the fan. Mild steel of 2mm thickness is selected.

Fan

It is made from aluminum due to its light weight. It has diameter of 30mm with length and thickness of 300mm and 2mm respectively.

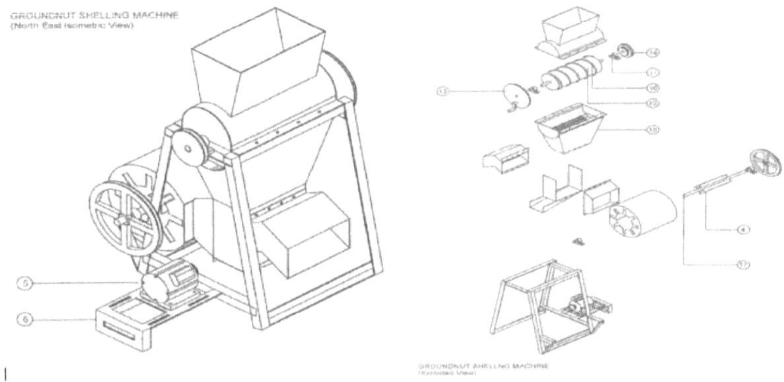

Figure 1: The Schematic Diagram of the Assembled Machine and Exploded View

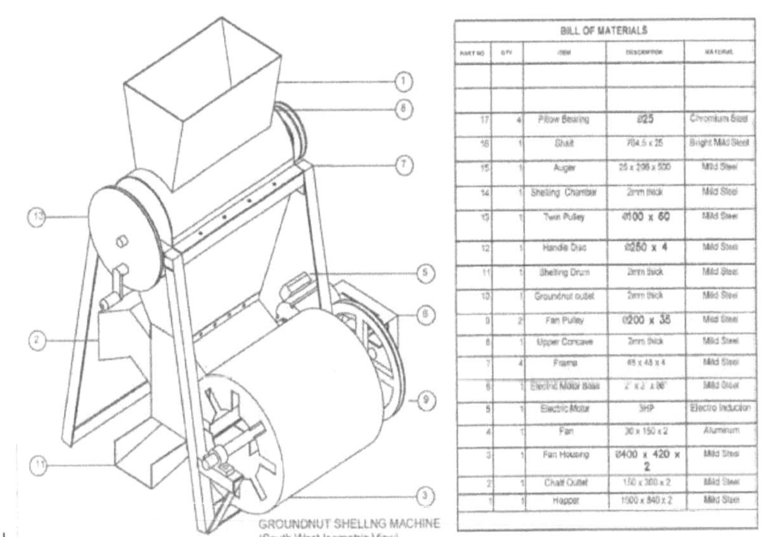

PART NO	QTY	ITEM	DESCRIPTION	MATERIAL
		BILL OF MATERIALS		
17	4	Pillow Bearing	Ø25	Chromium Steel
16	1	Shaft	RØ4.5 x 25	Bright Mild Steel
15	1	Auger	25 x 206 x 500	Mild Steel
14	1	Shelling Chamber	2mm thick	Mild Steel
13	1	Twin Pulley	Ø100 x 60	Mild Steel
12	1	Handle Disc	Ø250 x 4	Mild Steel
11	1	Shelling Drum	2mm thick	Mild Steel
10	1	Groundnut outlet	2mm thick	Mild Steel
9	2	Fan Pulley	Ø200 x 35	Mild Steel
8	1	Upper Concave	2mm thick	Mild Steel
7	4	Frame	45 x 45 x 4	Mild Steel
6	1	Electric Motor Base	2 x 2 x 90°	Mild Steel
5	1	Electric Motor	3HP	Electro Induction
4	1	Fan	30 x 150 x 2	Aluminum
3	1	Fan Housing	Ø400 x 420 x 2	Mild Steel
2	1	Chaff Outlet	150 x 300 x 2	Mild Steel
1	1	Hopper	1000 x 840 x 2	Mild Steel

Figure 2: The Assembled Machine and Part List

2.3 Design Analysis and Calculation

2.3.1 Determination of Crushing Power to Break the Pod

The crushing power (P) to break the pod as computed by Okegbile *et al.,* (2014):

$$Power = CLWF$$

(1)

Where, C = Convey or Capacity (m^2/s),

L = Convey or Length (w),

W = bulk material weight (N/m^3),

F = material factor.

Taken Cc = 800kg/min

$$\rightarrow c_c = \frac{800x10}{60} = 116.66 N/s.$$

But, $C = \frac{c_c}{w} = \frac{116.66}{199.3} = 0.59 m^3/s$

By substitution,

P = 40.83W.

Base on specification, 1hp motor was selected.

2.3.2 Determination of the Pulley Diameters

The diameter, D of the auger pulley may be determined from the relation:

$$N_1 N_2 = N_1 N_2$$

(2)

Where,

N_1 = speed of prime mover (1400rpm),

N_2 = speed of driven pulley (auger),

D_1 = diameter of prime mover pulley = 0.125m,

D_2 = diameter of the driven pulley (auger).

By substitution,

$$D_2 = \frac{N_1}{N_2} x \, D_1 = \frac{1440 \, x \, 0.125}{1800} = 0.1m$$

The diameter, D of the fan pulley may be determined from the relation:

$$N_3 D_3 = N_4 D_4$$

(3)

Where,

N_3 = speed of auger pulley,

D_3 = diameter of auger pulley,

N_4 = speed of fan pulley,

D_4 = diameter of fan pulley.

$$D_4 \frac{N_3}{N_4} x N_3 = \frac{1800 \, x \, 0.1}{900} = 0.2\text{m}$$

2.3.3 Determination of the length of the belt

The length, L_1 of thebelt between auger and motoris determined from the equation as given by Khurmi and Gupta (2005):

$$L_1 = 2c_1 + \frac{\pi(D_1 + D_2)}{2} + \frac{(D_2{}^2 - D_1{}^2)}{4c}$$

(4)

Where L_1= length of, c = distance between centres of pulley of motor and auger, D_1 = diameter of the motor pulley, D_2 = diameter of the auger pulley. Then, L_1 = 1640.47mm

The length, L_2 of the belt between auger and fan is determined from the relation:

$$L_2 = 2c_{min} + \frac{\pi(D_3 + D_4)}{2} + \frac{(D_4{}^2 - D_3{}^2)}{4c}$$

(5)

Where,

c_{min} =distance between centres of pulleys of fan and auger, $= 2(D_4 + D_3) = 600mm$

D_3 = Diameter of the auger pulley,

D_4 = Diameter of the fan pulley.

L_2 = 1671.3mm.

Hence, a type B – standard V-belt of $1694mm$ pitch length was selected for each of the belts, L_1 and L_2 for maximum power transmission.

2.3.4 Determination of Tension in the Belt

The tensions, T in the belt of length L_1 is shown below:

The belt speed V_b is estimated as given by Khurmin and Gupta (2005) and Akerele and Ejiko (2015);

$$V_b = \pi \frac{D x N}{60}$$

(6)

Where,

N = Angular speed of motor (rpm),

D = Diameter of motor pulley $V_b = 9.426m/s$.

The angle of contact,$\phi_1$for the belt around auger and motor (Hall *et al.*, 1983):

$$\phi1 = (180^0 - 2sin^{-1}(\tfrac{r_2-r_1}{c})x\,\tfrac{\pi}{180}$$

(7)

Where,

r_2 = radius of motor pulley,

r_1 = radius of auger pulley,

c = centre to centre distance between auger and motor,

$\phi1 = 3.103rad$

For a v-belt, the relationship between the tensions T$_1$ and T$_2$ in the belt is given by (Khurmi and Gupta, 2005):

$$2.3log\frac{T_1}{T_2} = \mu\phi_1 cosec\beta$$

(8)

Whereμ= coefficient of friction = 0.3,

ϕ_1 = angle of contact on smaller pulley,

β = half of groove angle = 19^0.

By substitution,

$T_1 = 17.507T_2$

(9)

Recall,

power, $P = (T_1 - T_2)V_b,$

where $P = 3.7285KW,$

V_b = belt velocity = 9.426.

Then,

$T_1 - T_2 = 395.55N$

(10)

By simplifying the equation2 9 and 10:

$T_1 = 419.506N$

and

$T_2 = 23.9621N$

Initial tension in the belt,

$T_1 = (T_2 + T_1)/2 = 221.73N$

The tensions, T in the belt of length L$_1$ is shown below:

The angle of contact,$\emptyset_2$ for the belt around auger and fan is given in equation 11 by Hall *et al.*, (1983) and Akerele and Ejiko, (2015)

$$\emptyset 1 = (180^0 - 2sin^{-1}(\frac{r_4-r_3}{c})x\frac{\pi}{180}$$

(11)

Where r_4 = radius of fan pulley, r_3 = radius of anger pulley, c_2 = centre distance of fan and auger pulleys.

$$\emptyset 2 = 2.9751 rad.$$

$$2.3 log\frac{T_3}{T_4} = \mu\emptyset_2 cosec\beta$$

(12)

$$T_3 = 15.5574 T_4 \ 13$$

Also,

$$power = (T_3 - T_4) V_b$$

(13)

$$\rightarrow T_3 - T_4 = 395.55$$

(14)

By simplifying equations 13 and 14:

$$T_3 = 422.7217N$$

And

$$T_4 = 27.1717N$$

Initial tension in the belt,

$$T_1 = (T_3 + T_4)/2 = 224.95N.$$

2.3.5. Determination of the bending moments acting on the shaft

The resultant of the tensions on belt 1,F_{r1};

$$F_{r1} = T_1 + T_2 = 419.506 + 23.9621 = 443.4681N$$

The resultant of the tensions on belt 2, F_{r2};

$$F_{r2} = T_3 + T_4(at\ 600) = 422.7217 + 27.1717 = 449.893N\ at\ 600$$

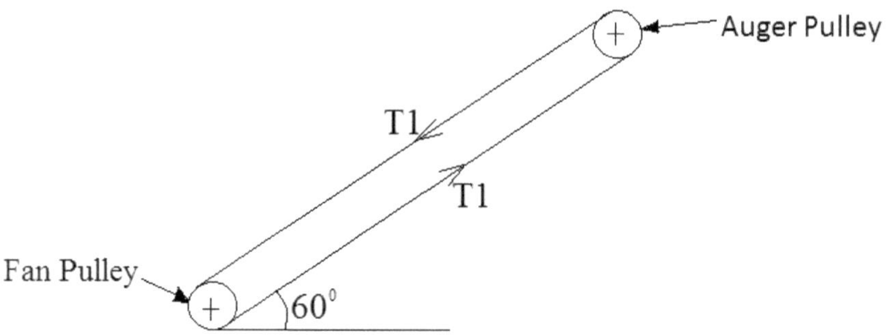

By resolving the forces into its vertical and horizontal components;

$F_{v2} = F_r 2 sin\emptyset = 449.8934 sin 600 = 389.6191N$

F_{H2} Vertical weight of the pulley, $wp = 9.81N$.

So, resultant vertical head of the pulley is given as:

$F_{r1} + F_{v2} + Wp = 443.4681 + 389.6191 + 9.81 = 842.8972N$

Wt. of auger, $w_1 = 0.15KN/m$,

Vol. capacity of hopper, $V = 0.5(a + b)\, hl = (0.25 + 0.15)\,0.25 \times 0.42 = 0.021m^3$

By experiment, the specific weight of dry groundnut is 199.3N/m³.

Therefore the hopper capacity $= 0.021 \times 199.3 = 4.1853N$.

Weight of the handle disc = volume of disc x specific weight of steel

$$= \frac{\pi x D^2 x h x 75}{4} = 0.003142 \times 75 = 236N$$

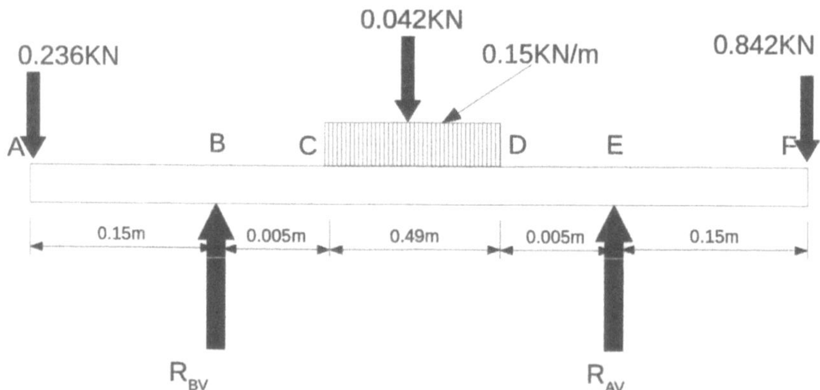

Figure 3: Vertical load diagram

Figure 4: Horizontal load diagram

Considering Fig 1: $\sum F_V = 0$,

$$\sum F_V = 0; \; R_{Av} + R_{Bv} = 0.236 + 0.042 + 0.15\,(0.49) + 0.842$$

$$(15)$$

$$\sum M_E = 0; \; 0.236\,(0.65) + 0.042\,(0.25) + 0.15\,(0.49)\,(0.25) = 0.5R_{Av} + 0.842(0.15)$$

$$(16)$$

By simplifying, $R_{Av} = 0.112\text{KN}$ and $R_{Bv} = 1.0815\text{KN}$

Using bending moment diagram; maximum bending moment, $M_{bmax} = 0.1263\text{KNm}$ at 0.15m from point F.

Considering Fig 2: $\sum F_V = 0$; $R_{AH} + R_{BH} = 0.224$ and $\sum M_E = 0$ $0.5R_{AH} = 0.15 \times 0.224$

(17)

By simplifying equation 17; $R_{AV} = 0.0672\text{KN}$ and $R_{BH} = 0.1568\text{KN}$

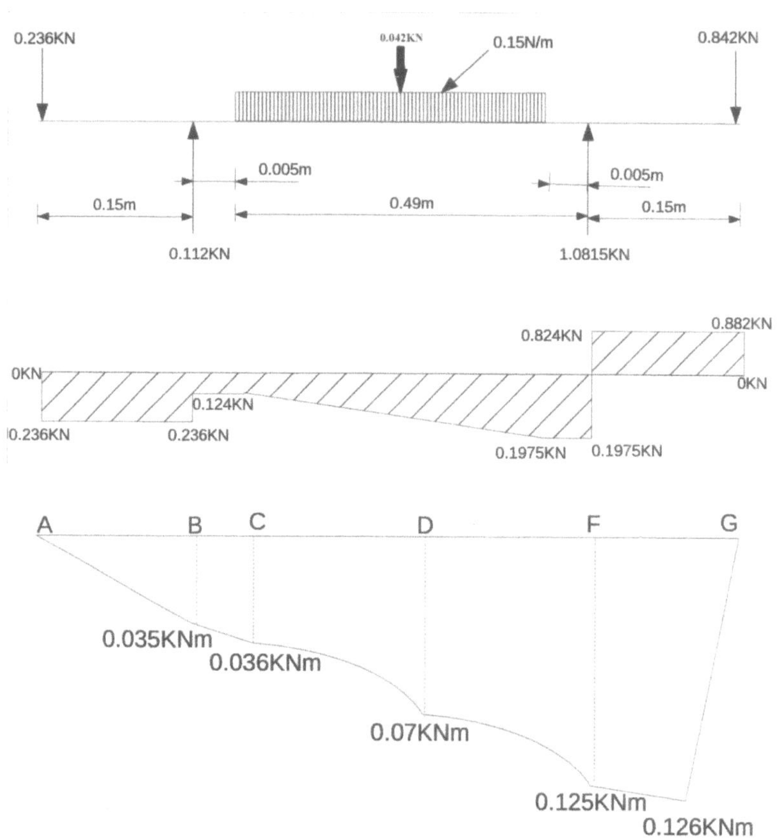

Figure 5: Free Body Diagram, Share Force Diagram and Bending Moment Diagram for Vertical Loading

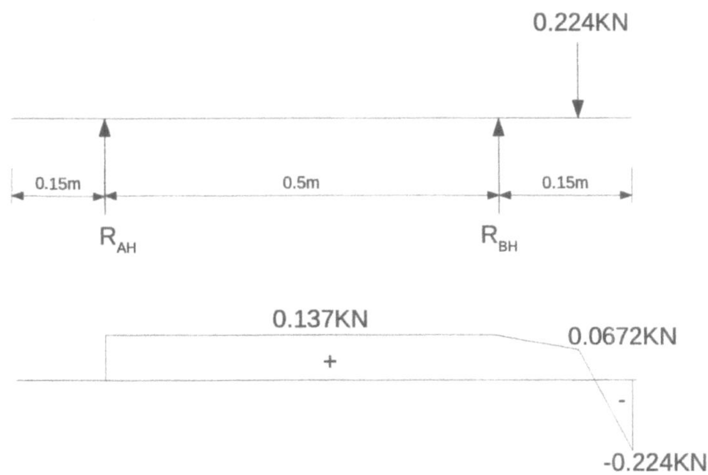

Figure 6: Free Body Diagram, Share Force Diagram and Bending Moment Diagram for Horizontal Loading

Considering points E for maximum resultant moment;

At E; $M_E = \sqrt{M_v{}^2 + M_H{}^2} = \sqrt{0.0336^2 + 0.1263^2} = 0.1307\text{KNm}$

Therefore, the maximum BM = 0.1307KNm which occurs at E.

2.3.6 Determination of the shaft diameter for auger

The shaft diameter, D is estimated using ASME code equation by Ejiko *et al.,* (2009) and Bhandari (2010)

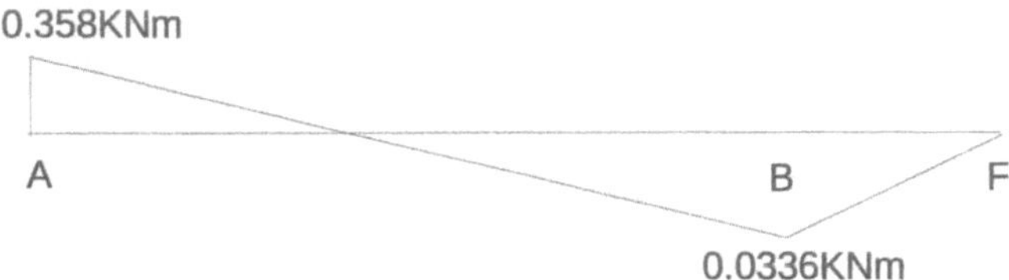

$$D^3 = \frac{16}{\pi x \tau_{max}} \left(\sqrt{(k_b x M_b)^2 + (k_t x M_t)^2} \right)$$

(18)

Where τ_{max} = permissible shear stress, M_b= maximum bending moment (BM), M_t = maximum twisting moment, k_b, k_t,= combined shock and fatigue factor applied to bending and twisting moment. Taken,

k_b = 2.0, k_t = 1.5, τ_{max} = 42mpa with Factor of safety (FOS) = 2.1

The maximum torque, M_t that a shaft can transmit may be determined from the relation:

$$M_t = (T_1 - T_2) R$$

(19)

Where T_1 and T_2 are tensions in the belt, R = radius of the auger shaft.

$$M_t = (419.506 - 23.9621) \times 0.1/2 = 19.7772Nm$$

So, $\mathrm{D}^3 = \dfrac{16}{\pi x42x10^6}\left(\sqrt{(130.7x2)^2 + (19.7772x1.5)^2}\right)$

$D = 0.0317m$. A standard size of $25mm$ was selected.

2.3.7 Determination of the shaft material

The selection of suitable material for the auger shaft is done as computed by Benham *et al.*, (1998):

Bending stress, $\sigma_b = \dfrac{My}{I}$

(20)

(Where M = bending moment, y = centroid of the shaft, I = moment of initial).

The critical part of the shaft is at $E.M = 0.1307KNm, y = 0.05m, I = \dfrac{\pi d^4}{64} = 4.909 \times 10^{-6}m^4$. Then, $\sigma_b = 1331.2kpa$

Torsional stress, $\tau_s = \dfrac{Tr}{j}$

(21)

Where T = torque, r = radius of shaft, j = polar moment of inertia. T = 19.77Nm, r = 0.05,

$j = \dfrac{\pi d^4}{32} = 100.67kpa$

The shear stress, $\tau_v = \dfrac{4V}{3A}$

(22)

(Where v = shear force, A = area of the shaft).

$v = 0.842^2 + 0.224^2$

$\rightarrow v = 0.8713KN.$

So, $\tau_v = \left(\dfrac{4x0.8713}{3x7.855x10^{-6}}\right) = 147.9kpa$. In order to estimate the principal stresses, we use the relation:

$$\sigma_{1,2} = \frac{\sigma_x + \sigma_y}{2} \pm \left(\frac{\sigma_x - \sigma_y}{2} + \tau_{xy}{}^2\right)^{0.5}$$

(23)

Where, $\sigma_x = \sigma_b$, $\tau_{xy} = \tau_v + \tau_s$. Then, $\sigma_1 = 1639.07$kpa, $\sigma_2 = 0$ kpa and $\sigma_3 = -307.95$kpa

Von mises equation is used to design against material failure as shown below (Rajput, 2006):

$$S_y = \sqrt[2]{\frac{((\sigma_1 - \sigma_2)^2) + ((\sigma_2 - \sigma_3)^2) + ((\sigma_1 - \sigma_3)^2)}{2}}$$

(24)

By substitution, Sy = 1.8128Mpa. A standard shaft made from mild steel is selected.

2.3.8 Determination of the size of power to drive the fan

The required power, P (watt) is obtained from the relation below:

$$\text{Efficiency, e} = \frac{\text{power output } (P_o)}{\text{power input } (P_i)}$$

(25)

Where p_o = fan flow rate (Q) x static pressure (P_s). Taken Q = 0.59m^3/s, P_s = 400N/m^2, e = 0.80

Then, Power input, $p_i = \frac{0.59 x 400}{0.80} = 295$W

2.3.9 Determination of auger capacity

The auger capacity, c is determined as given in the equation (Olumide, 1991)

$$C_1 = \frac{\pi x (D^2 - d^2) x P x N x P_b}{4}$$

(26)

Where D = screw auger diameter, d = shaft diameter (m), p = Screw pitch diameter (m), N = angular, speed (rod/s), P_b = bulk density of the material. C = 0.1m^3/s. So, the auger can contain about 420g of groundnut at a time.

2.3.10 Determination of a size of the key

The size of the key is determined from the relation as computed by Akerele and Ejiko, (2015):

$$\text{Width, } b = \frac{3D}{16} + \frac{25.4}{8}$$

$$\text{Thickness, } t = \frac{3D}{32} + \frac{25.4}{8}$$

Where D = diameter of the shaft (mm).the width and thickness are $7.86mm$ and $5.52mm$ respectively.

2.4   Fabrication Process

The fabrication was done at mechanical workshop located at Department of Mechanical Engineering, Federal Polytechnic, Ado-Ekiti, Ekiti State, Nigeria. It involved purchase of needed materials after thorough understanding of the design involved in the research. Several operations, such as marking out and cutting of components, bending and folding of sheet metals, welding, drilling, deburising by grinding, assembling and painting were done. There is re-designing and modification at certain stages of the fabrication until the whole machine was completed.

Figure 7: the assembled shelling machine

2.5   Operation of the shelling machine

The dried nuts were fed into the machine through the hopper. A prime mover (electric motor) is connected to the shelling and separating chambers with the aid of compound pulley. As it supplies the power, the auger shaft rotates and breaks the pods against the shelling chamber wall and stationary sieve. The air produced by the action of the fan (or blower) sweeps the seeds and shell fragment out of the shelling zone into the inclined outlet unit. The fan passage serves as the separation chamber such that the seeds are collected at the discharge outlet while the pods are collected at the shaft outlet.

2.6   Performance Evaluation

The test was carried out with a 5 horse power (1440rpm) electric motor using groundnut seeds of various quantities. The unshelled groundnut were dried. Then, the machine was fed with a constant mass of 300g of groundnut seeds. Upon shelling, the number of seeds shelled and unshelled were counted separately and weighed. The efficiency is obtained from the relations shown below:

$$\text{Throughput capacity} = \frac{\text{Quantity of groundnut fed}}{\text{Time taken}}$$

$$\text{Shelling efficiency} = \frac{\text{weight of groundnut shelled}}{\text{Weight of groundnut fed into machine}} \times 100\%$$

$$\text{Moisture content} = \frac{\text{weight of water}}{\text{Weight of wet groundnut}} \times 100\%$$

$$\text{Materials damage} = \frac{\text{weight of broken seeds}}{\text{Weight of groundnut fed into machine}} \times 100\%$$

(Is 320 – 1997: Nigeria Standard Test Code for Power Sheller's, Grains and Seed Nis. 320 (1997) Nigeria Industrial Standard Test Code for Grain and Seed Cleaners)

Table 1: Result of shelling efficiency

Test	Total weight (g)	Shelled (g)	Broken (g)	Unshelled (g)	% shelled	% broken
I	300	249	40	1	83	13
II	300	251	44	5	84	15

Table 2: Result of average shelling efficiency

	Average shelled	Average broken
% by mass	84	14

Table 3: Result of moisture content

Test	Initial weight	Final weight	Weight loss	% moisture loss
1	6666	900	5766	86.5

3.0 RESULT AND DISCUSSION

Table 1 shows the data collected for the shelling efficiency of the machine. Table 2 shows the calculated average shelling efficiency. The result of test conducted on the percentage dryness is presented on Table 3. It was deduced from the test data that the groundnut seeds used are 86.5% dry. The machine is having a shelling efficiency of 84% and material damage of 14%. The assembled machine is shown in Figure 5 and the exploded view is shown in Figure 6.

4.0 CONCLUSION AND RECOMMENDATIONS

A groundnut shelling machine has been designed, fabricated and tested in this research. A preliminary test evaluation in terms of shelling efficiency and material damaged has indicated that it has a higher potential in substituting manual methods. Also, the machine exceeded the previously designed and fabricated shelling machine in terms of efficiency and time. This is because the design involves material strength and rigidity. The following recommendations are required for effective utilization of the machine; these include making sure the groundnut moisture content is not more than 16%, running the machine for a maximum of 10 hours daily, installing the machine in a well ventilated area, using engine grade lubrication oil and running daily maintenance after operation to prolong the machine life span.

REFERENCES

Akerele O. V. and Ejiko, S.O. (2015), "Design and Construction of Oil Expeller", International   Journal Of Engineering And Computer Science, India, ISSN 2319-7242 Vol. 4, Issue 6,   pp. 12529 -12538.

Benham, P.P., Crawford, R.J. and Armstrong, C.G. (1998). "Mechanics of Engineering material", 2nd ed. Longman publisher Ltd., Singapore, pp 301, 341-344

Bhandari, V.B. (2010). "Design of Machine Elements", Tata McGrawHill Education, pp 330-   334.

Ejiko S. O., Adelegan G. O. and Dirisu N. O. (2009): "Development and Performance Evaluation of a wear Testing Equipment" Journal of Engineering & earth Sciences (JEES) Vol 4, No 2 pp 6-11

Khurmi, R.S. and Gupta J.K. (2005). "A Textbook of Machine Design", multicolour illustrative ed., Eurasia publishing house (PVT) Ltd. by S. Chand & Co. Ltd, India.

Komolafe, M.F., Adegbola, L.A and Asgate, T.I. (1985). "Agricultural Science for West African School and Colleges", 2nd ed., University press Ltd., Ibadan

Hall, A.S., Holowenko, A.R and Laughlin, A.G. **(1983)**. "Theory and Problems Of Machine Design", Schaum's outline series, McGrawHill Book co. NY.

Hommons, R.O. (1994). "The origin and history of groundnut In: The groundnut crop; A scientific basis for improvement": J. Smart, 1st ed. Champman and Hall, NY, pp56-78

Maduako, J.N. and Hamman, M. (2004). "Determination of Physical Properties of Three Groundnut Varieties". Nigerian Journal of Technology, Vol. 24(2), pp 12-18.

Maduako, J.N., Saidu, RA, Matthias, P. and Vanke, I. (2006). "Testing of an Engine Powered Groundnut Shelling Machine". Journal of Agricultural Engineering and Technology, vol. 14, pp

Milner, H.G (1973). "Traditional Extraction of Oil from Nigerian Food Stuff". Heinemann Education Books, Ibadan.

Okegbile, O.J, Hassan, A.B, Mohammed, A and Obajulu, O. (2014)."Design of a combined Groundnut Roaster and Oil Expeller Machine", International journal of science and Engineering Investigations, vol. 3, issue 26, pp 26-30

Olumide A.A. (1991). "Design and Experimentation Of Auger screws". Unpublished Bsc Thesis, Department of Mechanical Engineering, Obafemi Awolowo University.

Rajput, R.K. (2006). Strength of Materials, Multicolor Illustrative ed., S.chand & co.Ltd., New delhi

Rodriquez, M and Carruthers, I. (1985). Tools for Agriculture: A buyers guide to appropriate equipment, 3rd ed. IT publications, London, pp 99,100,117

Shankarappa T., Robert, E.R and Virginia, N. (2003). World Geography of groundnut: distribution, production, use and trade.

YOUR KNOWLEDGE HAS VALUE

- We will publish your bachelor's and
  master's thesis, essays and papers

- Your own eBook and book -
  sold worldwide in all relevant shops

- Earn money with each sale

Upload your text at www.GRIN.com
and publish for free